Primary Science

for the Caribbean

Tony Russell

Consultant Editors:

Adrian Mandara
Achisha Jaikaransingh

Book 6

Nelson Thornes Primary Science

Nelson Thornes

Published in 2004 by:
Nelson Thornes Ltd
Delta Place
27 Bath Road
CHELTENHAM
GL53 7TH
United Kingdom

13 / 15 14 13 12 11 10

A catalogue record for this book is available from the British Library

ISBN 978 0 7487 7806 5

Illustrations by Jane Bottomly

Page make-up by IFA Design Ltd

Printed in China by 1010 Printing International Ltd

Primary Science for the Caribbean Teacher Guide Books 4–6 is available online at www.nelsonthornes.com and in hard copy form from your local agent

Acknowledgements

Photo Credits. Alamy: Laura Dwight/Stock Connections p. 45; Axon Images: p. 53; Bruce Coleman: Pacific Stock p. 65 (left); Corbis: Douglas Peebles p. 32 (a), p. 53; Corel (NT): p. 13 (a), (b), (c), (d), (f), p. 26, p. 32 (e), p. 36 (right), p. 37, 58; Digital Vision (NT): p. 13 (e), p.57 (bottom right); Eye Ubiquitous: Philip Wolmuth/Hutchison p. 31; Frank Lane Picture Library: John Holmes p. 57 (top left), E & D Hosking p. 65 (left); Holt Studios International: (second from left, middle); ICI: p. 57 (top right); Martyn Chillmaid: p. 18, p. 26 (centre); Mediscan: p. 47 (bottom right); Panos Pictures: Giacomo Pirozzi p. 47 (top left), Crispin Hughes p. 47 (bottom left); Photodisc6 (NT): p. 26; Rex Features: p. 51; Science Photo Library: p. 50 (left), Adam Hart-Davis p. 7, Dr Jeremy Burgess p. 36 (bottom left), Kent Wood p. 50 (second from left), Eye of Science p. 50 (second from right), Dr Tony Brain p. 50 (right), Dr P. Marazzi p. 50 (bottom), Michael P. Gadomski p. 53, Novosti p. 59; Still Pictures: p. 45 (right), p. 62 (top right); Stockbyte (NT): p. 26; Trip: Helene Rogers p. 62 (top left, bottom), p. 64; Wellcome Institute Picture Library: p. 17, p. 47 (top right); www.tropix.co.uk: V & M Birley p. 32 (second from right).

Book 6 Contents

Sense organs: human eyes and ears

The structure and function of human eyes and ears

Activity 1

1 Look closely at the eyes of your partner, but DO NOT TOUCH them.

2 Record your observations in notes and drawings. Note how the eyes function, as well as their structure.

3 Compare your observations with those of your partner who has observed your eyes. Note any differences and similarities.

4 Share your observations with the class and discuss what the class has found out about the structure and functioning of the eyes.

You will need: paper and a pencil.

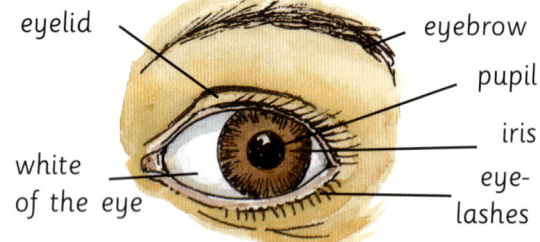

eyelid • eyebrow • pupil • iris • white of the eye • eye-lashes

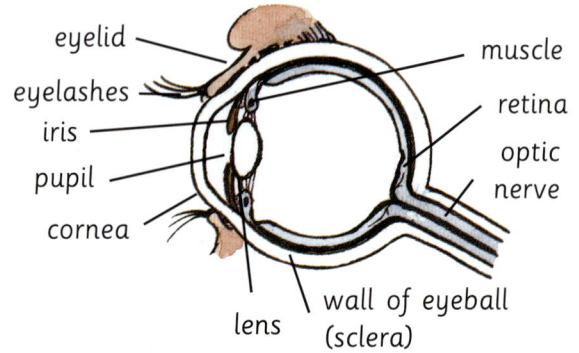

eyelid • muscle • eyelashes • retina • iris • optic nerve • pupil • cornea • lens • wall of eyeball (sclera)

The human eye is about 2.5 cm in diameter and is roughly spherical. It fits into the eye socket, a space in the front of the skull. It has six muscles attached to it, which can move it in various directions. A tear gland in the top of the eye produces tears that keep the eyeball moist and lubricated. The tears drain from the eye into the nose.

Activity 2

You will need: model of the eye, videos, CD-ROMs, books that contain information about the eye, an animal's eye, dish, alcohol or other preservative, jar with lid, paper and a pencil, tweezers or mounted needles.

1. Use the model and other resources to understand the structure of the eye. Make sure that you understand where each part is located and how it is connected to others.

2. Draw a diagram of a cross-section through the eye, labelling all the parts shown in the diagram in Activity 1.

3. Carefully observe the external features of the animal's eye and compare it with the human eye. Handle the eye with tweezers or mounted needles, rather than handling it directly. Make notes and drawings of your observations.

4. Put the animal's eye in a jar filled with alcohol or other preservative and put the lid on tightly.

5. Use the resources to find out how the eye functions. Make notes on each of the parts listed: cornea, lens, pupil, iris, retina, optic nerve, muscles.

6. Share your findings with the class and listen to what others tell you. Ask questions about anything that is not clear about the functioning of the eye.

Vision (eyesight) is probably the most important human sense. A large part of the brain is devoted to vision. The front of the eye is covered with a transparent cornea, which bulges out slightly. It protects the more delicate parts of the eye, which are behind it. Light enters the eye through the pupil. This appears as the black, circular centre of the eye when we look at it directly from the front. The size of this hole can be changed, depending on the amount (intensity) of light arriving at the eye.

Activity 3

You will need: paper and a pencil.

1. Sit facing your partner and decide who will do the observation first.

2. The one being observed should look towards the light coming into the room through a window. You should look closely at their eyes and note the iris and pupil. Do NOT tell your partner what you see.

3. Your partner should now close their eyes and keep them closed for one minute.

4. When the minute has passed, tell them to open their eyes and look again at the light. You must be quick to observe what happens. Do NOT tell your partner what you see. Make a note of your observations.

5. Now repeat the activity with you closing and opening your eyes, as your partner observes.

6. Tell one another what you observed and compare your observations.

7. Discuss what you saw and try to explain it. Write down your explanation.

8. Tell the class what you think.

The iris is the coloured ring of tissue that surrounds the pupil. The size of the pupil is changed by the contraction and relaxation of the iris. This is brought about by tiny muscles in the iris. When the light is strong, the iris automatically relaxes and closes down the size of the pupil to a minimum. When the level of light is low, such as at night, the iris automatically contracts and opens the pupil to its maximum size. This is 16 times bigger than its minimum size. We do not have to think about the size of our pupils, as they react automatically to the levels of light and adjust their size accordingly.

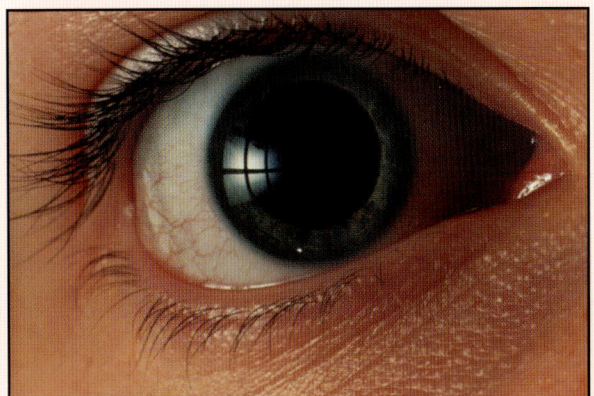 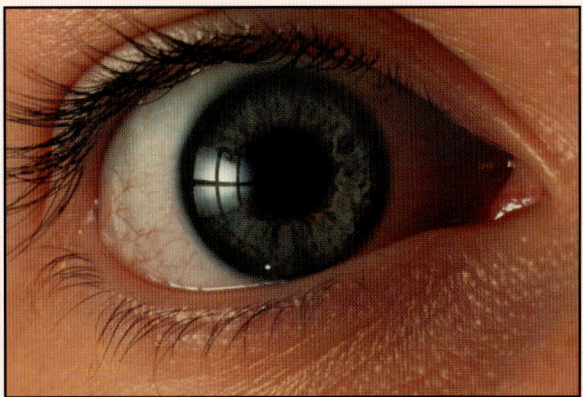

The eyelashes help to keep dust and other solid particles out of the eyes. As the eyelids blink repeatedly they act like windscreen wipers on a car. They push the tears over the eyes and so wash its surface clean. The tears, plus whatever particles may be trapped in them, are drained into the nose through small openings at the inside corners of the eyes. When we blow our noses, this material, along with dust, etc. trapped by the mucus and hairs in the nose, is pushed out of our bodies.

Activity 4

1 Explore how the lens interacts with the light from the torch or other source to produce an image on the sheet of paper. This will be done best in a darkened room or 'blacked-out' area of the room.

You will need: a convex lens, a torch or other light source, paper and a pencil.

2 Arrange the three items so that you see the best possible image on the paper. Observe the image carefully and record what you see in a drawing. Try to explain in writing what you have drawn.

3 Share your results with the class.

Discuss with the class how these observations are related to the working of the eye.

The inner surface of the eye is covered by the light-sensitive retina. This thin layer contains cells that react to light. When the chemicals in the cells are stimulated by light, they produce electrical impulses that travel along nerve fibres in the retina, then away from the eye along the optic nerve to the brain. Some of the cells in the retina respond to the colour of the light (cone cells) and others only respond to the intensity of the light (rod cells). The optic nerve transmits the impulses to the part of the brain that can interpret them. The eye does not know or understand what it is 'seeing': it is the brain that 'makes sense' of the images formed on the retina.

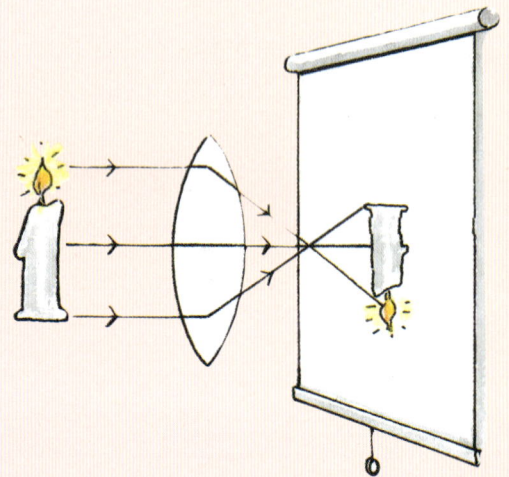

When light passes through the lens, which is situated behind the pupil, it has its direction changed by the lens. This focuses the light on the retina, so that a clear image can be formed. The lens inverts the image – it turns it upside down. It also reverses it from left to right. It is the brain that corrects these changes, so that we 'see' the world the right way round!

The lens is not hard, like glass. It is elastic and can change its shape, so that we can look at things far away and things that are close. The lens is attached to tiny muscles which can contract and relax, just like the iris. When they contract, they

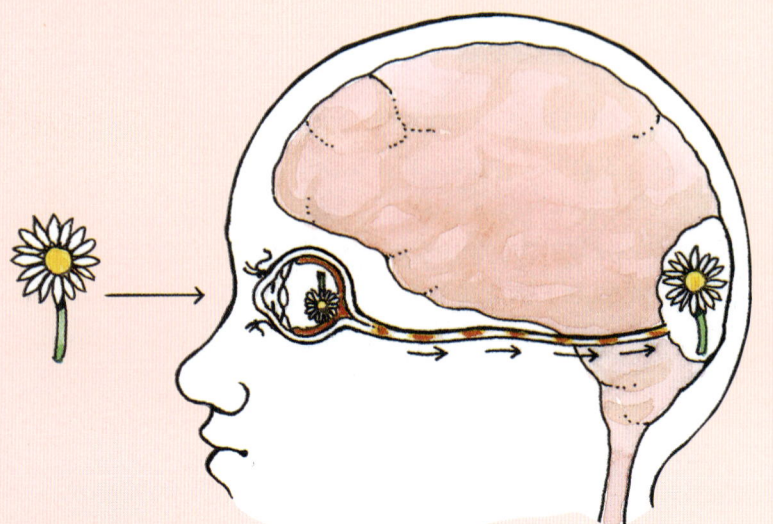

pull on the lens and it is made thinner. This is best for seeing distant objects. When the muscles relax, the lens returns to a more curved, thicker shape, which is best for seeing nearby objects. The shape of the lens is always suited to the production of clear, focused images on the retina.

Activity 5

You will need: materials for model making, or a large sheet of paper and colouring materials for poster/chart making.

1. Discuss with your group what form your product will take – a 3D model of an eye, or a chart/poster illustrating the structure and function of the eye.

2. Collect the materials you need for your chosen product. Share out the tasks involved in making it.

3. Make the parts and assemble them to make the complete product. Display it for the class to see.

4. Use it to explain to the class how the eye produces an image of what it is looking at.

Humans, like most other animals, have two eyes. The images formed in the two eyes are not identical. The field of vision of each eye overlaps with the other, so that images of objects directly in front are seen most clearly, and those to one side less clearly. The brain uses the two slightly different images to produce 'depth' in our view of the world. This is essential for us to judge distance and movement. The two optic nerves cross over one another on their paths to the back of the brain, where our visual centres are located. So, the images from the right eye are received and interpreted by the left side of the brain and those from the left eye are processed by the right side of the brain.

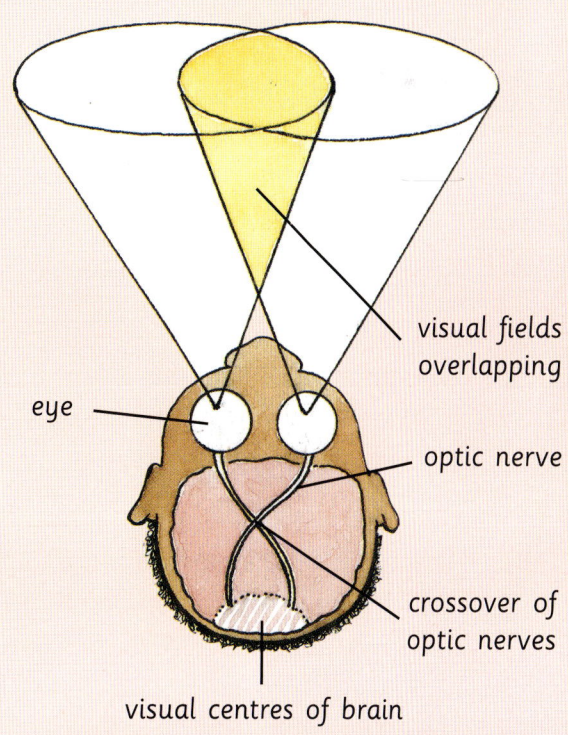

visual fields overlapping

eye

optic nerve

crossover of optic nerves

visual centres of brain

Activity 6

1 Blindfold one of your group. Put the tray of small objects in front of them and tell them that they have to pick up each one as quickly as they can.

You will need: various small objects, a tray, a cloth for a blindfold, a timer or watch, paper and a pencil, books and other resources about visually challenged people.

2 Draw a table for keeping a record of the results. Have columns for 'blindfold', 'one eye' and 'two eyes'.

3 At the same time as you tell them to begin, start the timer. When they have collected all the objects, note down how long it took.

4 Now uncover one eye and repeat the activity.

5 Lastly, uncover both eyes and let them pick up the objects again.

6 All the others in the group should carry out the same three steps and their times should be recorded in the table.

7 Compare how long it took each time. Use your knowledge of how the eyes work to explain the differences.

8 Share your results with the class.

9 Discuss how individuals adapt to losing the sight of one or both eyes. Use the books and other resources to research the ways in which visually challenged individuals adapt their way of living. Make notes of what you find out.

10 Use the resources to find out about the causes of people losing their sight and how to prevent it happening through carelessness or accidents. Make notes of what you find out and share them with the class.

Look at the pictures and discuss with your group how the things are connected with looking after our eyes and keeping good eyesight.

Make a list of all the things we can do to keep our eyesight safe. Share it with the class.

Activity 7

You will need: mirrors, lenses, paper and a pencil.

1 Use the mirrors and lenses to explore how the eyes can be misled. Try to set them up so that children from another group will be 'tricked' by what they see.

2 Ask another group to look at what you have done. They should try to explain how you have tried to 'trick' them.

3 Now look at what the other group has done and try to work out how it was done.

4 Write a simple explanation of what your group did with the mirrors or lenses.

5 Tell the class about times when your eyes have misled you. Try to explain how it happened. Listen to what others tell you.

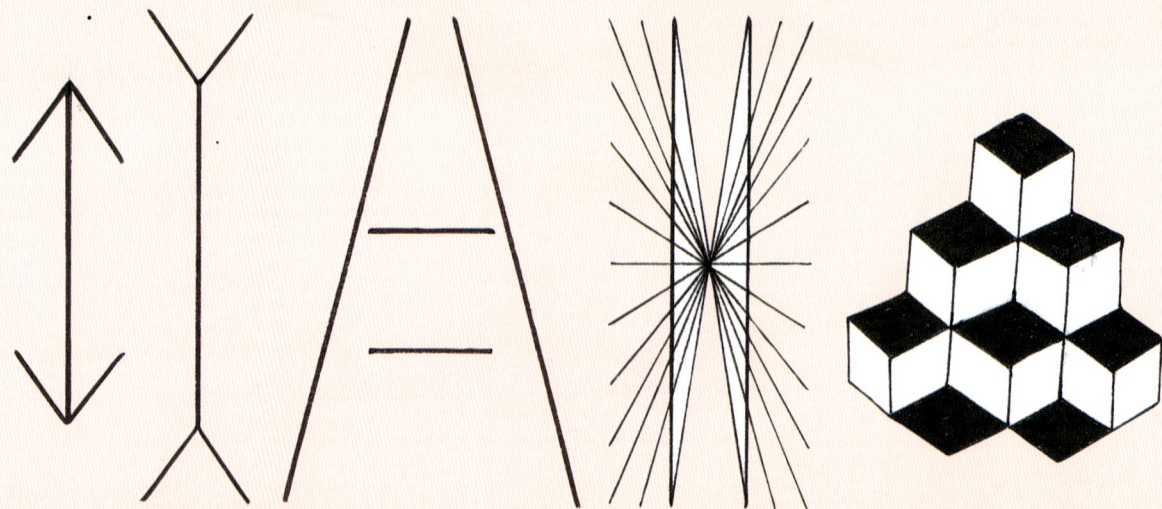

Look at the drawings and tell your partner what you see in each one. Let them tell you what they see.

If there are differences, try to work out why.

If you are not sure what your eyes are 'telling' you, use a ruler to check some of the drawings.

These drawings are examples of optical illusions. They appear to be one thing, but really they are something else. Our eyes seem to be 'tricked' by them. They are a good way of discovering that our sense of vision has limitations. It is our mind, rather than our eyes, which has problems with these images.

The mirage, shown in this photograph, is another example of our eyes being misled. Objects, especially water, 'appear' in the distance, but they are not really there. The image is the result of

hot air changing the way light travels, so that our eyes cannot tell what is real and what is not.

Mirages are very common in hot dry places, such as deserts.

Optical illusions and mirages are examples of misperceptions. Our eyes send images to our brain, but they are not correctly understood. There is an error in our perception.

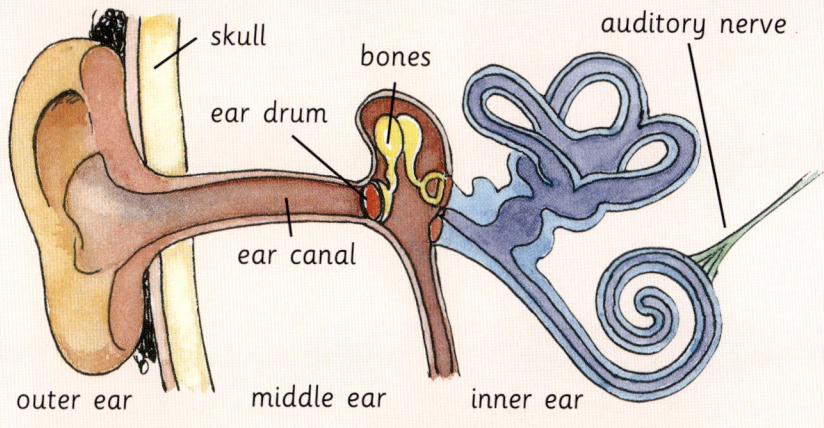

skull
bones
ear drum
auditory nerve
ear canal
outer ear
middle ear
inner ear

Look at the pictures of other animals that are mammals, like ourselves. Compare their ears with yours. Make notes of any differences you see or know about from your own observations. Discuss your notes with your group.

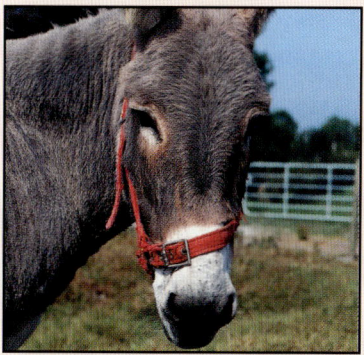

Activity 8

You will need: books and other resources that have information about the ears and how they do their work, paper and a pencil.

1 Draw a diagram of the ear with labels on all the parts shown in the picture above.

2 Use the books and other resources to find out how the ears work. Write simple explanations of how sounds are heard. Try to describe the function of each part of the ear.

3 Add notes about the part played by the brain in hearing and recognising sounds.

Activity 9

1 Use the books and other resources to find out all you can about the sense of hearing in different animals. You should choose which animals to do your research about.

You will need: books and other resources with information about the sense of hearing in animals, paper and a pencil.

2 Make notes about their hearing. Try to find out if it is the same or different in any way from our own sense.

3 Add notes about the part played by the brain in hearing and recognising sounds.

4 Write your conclusion about the hearing of animals compared with that of humans.

Whales have no ear holes or outer ears, but their sense of hearing is better than ours. Sound travels better through water than through air, so they can detect sound through the skin that covers their middle and inner ears. They do not need outer ears at all.

Many bats rely on their sense of hearing to fly about in the dark without hitting against things. They also use it to detect and to catch their food. They eat flying insects, such as moths, which they catch as they fly. The ears of the bat are very sensitive and are able to hear sounds at very high pitch, which we cannot hear at all. They make these high pitched sounds as they fly and their ears can hear the echo of the sounds, which are produced by objects hit by the sound waves.

Dogs can also hear sounds that we cannot hear. Their ears are able to detect high pitched sounds. There are special whistles that can be blown to call a dog. The person blowing the whistle cannot hear anything!

Cats and many other mammals can move their ears to help them locate the source of sounds. This is very useful when they are trying to catch their prey, as the slightest sound can tell the animal where the prey is hiding. Most people can only move their ears a very small amount – certainly not like a cat!

Elephants use their ears for cooling themselves, as well as for hearing. The large outer ears are flaps that can be moved backwards and forwards. They contain many blood vessels and the blood is cooled down as the ears flap. It is even more cooling if the ears are wet.

This table shows the range of hearing of humans and other animals.

Animal	Range of frequency (pitch) (Hertz)
Human	20–20,000
Cat	100–32,000
Dog	40–46,000
Elephant	16–12,000
Bat	1,000–150,000
Whale	70–150,000
Locust	100–50,000

Activity 10

1. Discuss with your group what you will present about how to care for our hearing. It should be done as a performance.

   You will need: items for a performance piece on taking care of our hearing.

2. Decide what form of performance it will be. Collect the necessary materials and prepare your piece. Rehearse it until you are ready to share it with the class.

3. Perform your piece, then ask the class what they think message was. Answer any questions they might have.

4. Watch the performances of other groups and ask them questions about anything that you do not understand.

Our hearing can be damaged if we do not care for it. The ear drum is easily torn or broken by hard, sharp objects. Such things should never be pushed into the ears.

Dirt and germs get into the ear canals – we cannot stop this. Our ears make wax, which helps to keep the ear drum flexible and able to vibrate. This wax, with the dirt and germs that get into the ears, can block them and it can lead to infections. If this gets into the middle or inner ear, the damage can be serious as well as very painful. Regular, gentle washing of the ears can help to remove the wax and dirt, so that the risk of infections is reduced.

Loud sounds, including music, can also damage our hearing. If we listen to such loud sounds day after day, our hearing is gradually damaged, little by little. It is the volume of the sound, i.e. the amount of kinetic energy, which is measured in decibels (dB). The more energy, the more damage the sound can do.

Activity 11

You will need: a blindfold, various objects that can be used to make sounds, paper and a pencil.

1. Sit in a circle and put one person in the centre of the circle. Blindfold them.

2. When everyone is silent, use one of the objects to make a sound. Ask the blindfolded person to name the object. Keep a record of their answer.

3. When they have listened to all the sounds, let another person sit in the centre, blindfolded and repeat the test. Keep a record of their answers.

4. When everyone has had their turn, compare the answers given by different people. Look for sounds which everyone named correctly and those which no one recognised. Discuss with the class why you think this happened. Compare the girls' answers with the boys' answers. What do you find?

5. Discuss how people whose hearing is limited or completely lost (aurally challenged), manage to live by adapting.

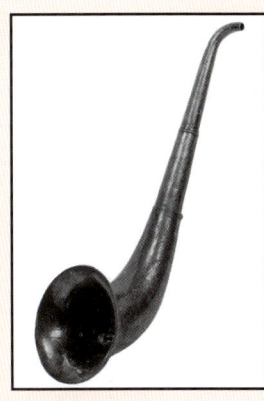

a) an ear trumpet

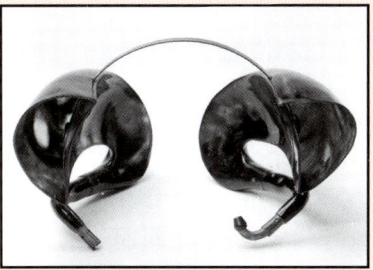

b) an early electric hearing aid

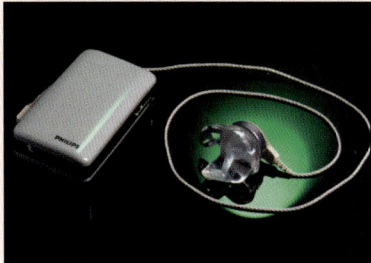

c) a modern electronic hearing aid

These pictures show some of the ways in which people have used science to help them overcome their hearing problems. The technology has improved a great deal since the time when the ear trumpet was the best thing available. Hearing aids have gradually become smaller and easier to use. This is good, because some people with hearing problems are very concerned about how other people treat them.

Activity 12

1. Write the two headings 'pleasant' and 'unpleasant' at the top of your paper.

   You will need: a tape player, a tape of different sounds, paper and a pencil.

2. Listen to each sound on the tape and decide whether it is pleasant or unpleasant for you. Write down its name in your chosen list.

3. When you have listened to all the sounds, share your lists with your group and explain your reasons for the way you sorted the sounds. What makes a sound unpleasant for you?

4. Collect as many examples of noise pollution as the group can think of. Share them with the class. Try to agree on what it is about each sound which makes it 'pollution'. Make a list of the features of polluting sounds.

5. Suggest ways in which the noise pollution can be reduced or removed completely.

Sounds are vibrations of the air caused by some vibrating object. They travel in all directions away from their source. If the vibrations reach our ears, the air in the ear canals begins to vibrate. The outer ear in humans is not important for our sense of hearing. It is not large and it cannot be moved to collect more of the sound.

At the bottom of each ear canal is a thin, flexible ear drum. When the vibrating air hits the ear drum it also begins to vibrate. The inside surface of the ear drum pushes against a set of three small bones, which amplify (increase) the vibrations and pass them on to the inner ear. The part of the ear where the three bones are is called the middle ear.

The inner ear has two important organs. One is our organ of balance and has nothing to do with hearing. The other is the sense organ for hearing. It is a coiled tube filled with liquid. Its walls have many nerve cells, each one of which is sensitive to sounds of a particular pitch. The liquid in the tube responds to the vibrating bones and the vibrations travel through the liquid and stimulate the nerve endings. The nerves send their signals along the auditory nerve to the part of the brain that can interpret the signals as sounds.

The transmission of light and sound

The_____is_____

The_____is_____

The_____is_____

Activity 13

(1) Draw a table with three columns. Head them **transparent**, **translucent** and **opaque**.

(2) Test each material one at a time to find out which set it belongs to. Be careful to make the test fair each time. Keep the light source the same distance from each of the materials as you test them. Write the name of each material in the correct column.

You will need: materials of various kinds, a light source, paper and a pencil.

(3) Share your results with the class. Discuss what you have all found out about materials.

Transparent materials allow almost all the light to pass through them. This means that we can see through them clearly. Water, glass, air, some plastic sheets and many other materials are transparent.

Translucent materials allow only a little light to pass through them, or they scatter the light as it goes through. This means that no clear image can be seen through it, though we may see the shapes of objects. Many plastic sheets, paper, china plates, clouds, some fabrics and many other materials are translucent.

Opaque materials do not allow light to pass through them. This means that we cannot see through them at all. Metals, wood, rocks, people, rubber, bananas, many fabrics and plastics are opaque. Even materials like paper, which are translucent, can be opaque if there are many layers one on top of the other. So, a page is translucent, but a book is opaque. Opaque objects block the light and this creates shadows.

light source · shadow · opaque object

Activity 14

You will need: a light source such as a window or lamp, a white sheet with pegs or pins to keep it up, paper and a pencil.

1 Set up the sheet as a screen, with the light behind it.

2 Discuss with your group a short story that you can act as a shadow puppet play.

3 Practise the play before you go behind the sheet to perform it for the class.

4 When the play is ready, go behind the sheet and use the light to make the shadows for the play.

5 Watch the shadow plays performed by other groups.

6 Write a short explanation of how the shadows are formed.

Activity 15

You will need: a straw or other flexible tube (e.g. a short length of garden hose), paper and a pencil.

1 Use the tube to look at an object. Record what you observe.

2 Bend the straw slightly so that it is not completely straight.

3 Use the tube to look at the same object again.

4 Record what you observe.

5 Discuss the result with your group and try to explain it.

6 Share the group's results and explanations with the class.

Light travels in straight lines from its source. This means that, if anything opaque is in the path of the light, it will be blocked. The light cannot 'go round' it and carry on its way. In a straight tube, such as a straw, the light rays travel along it without being blocked. That is why we can see through the straight tube. The image of the object we are looking at travels through the tube to our eye.

If the tube is bent or curved, there is no straight pathway for the light and so it cannot travel through the tube. We cannot see the object through the tube because the light is blocked by the bend or curve.

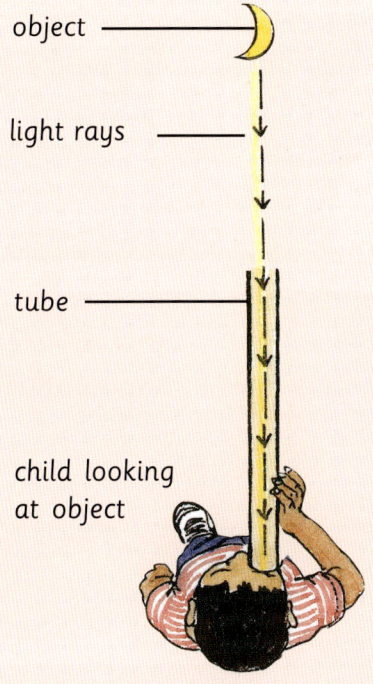

object

light rays

tube

child looking
at object

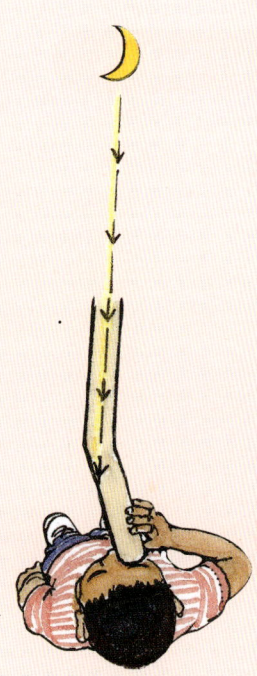

Activity 16

1. Explore the images you can see in the mirror when it is flat. Use it to look at yourself from different positions. Use drawings, diagrams or notes to record what you see.

You will need: a flexible mirror, paper and a pencil.

2. Carefully bend the mirror so that it is curved, either inwards or outwards. Use it again to look at yourself. Compare the images which you saw when it was flat. Record what you see this time.

3 Now bend the mirror the opposite way and repeat your observations. Compare the images which you now see with those which you recorded before. Record the new images in some way.

4 Discuss your observations with the group and try to explain them. Remember that light travels in straight lines.

When light hits surfaces it is more or less reflected. Some surfaces are bad reflectors, because they are rough and uneven. The light is scattered by the surface and travels away in all directions. We cannot see an image in these surfaces, e.g. the soil, a tree trunk, a pineapple, a car tyre. They do reflect light, but they would not be called reflectors, or reflective surfaces.

Other surfaces are good reflectors and we can see images reflected from them. Such surfaces are shiny and smooth, which means that light hitting them is reflected in an orderly way. It is not randomly scattered. So we see a reflected image of the object coming from the shiny surface. This is what a mirror does.

Activity 17

You will need: a light source such as a candle or electric lamp.

1 Light the candle or switch on the lamp.

2 Form a circle around the light source and look towards it.

3 Slowly move back, further and further from the source. As you do this, observe the light source and note what you see.

4 Keep moving away until you can go no further. Return to your seat and tell the class what you observed as you went further and further from the light source. Try to explain your observations.

Look at the pictures and sort the objects into two sets. One set should be luminous objects. The other set should be the non-luminous/illuminated objects. Write down your lists and share them with the class.

Activity 18

You will need: books and other resources about light sources and reflection.

1 Use the resources to find out more about luminous and non-luminous objects.

3 Display your work and look at what others have produced.

4 Discuss what has been found out by the class.

2 Make drawings and notes about what you find.

Luminous objects are sometimes called primary sources of light, because they make light that travels away from them. The sun, a candle, a fire and a firefly are examples of primary sources of light.

Non-luminous objects do not make light, but they do reflect it. They are sometimes called secondary sources of light. They seem to be making light, but really the light has come from a luminous object and is only reflected by the non-luminous object. The moon is a good example. It seems to be making 'moonlight', but actually it is only reflecting light from the sun. The moon is like a giant mirror. Most things around us are non-luminous. Some are good reflectors – such as the surface of the sea, a road sign that 'lights up' as headlights shine on it and a mirror. Most things are not such good reflectors, but even this page is reflecting the light in the room towards your eyes. If it did not, you would not be able to see it!

Activity 19

You will need: a transparent container for water, a pencil or stick, water, paper and a pencil.

1 Put the stick or pencil into the empty container and observe it from the side. Draw a picture of what you see. Look at it from above and make another drawing.

2 Half-fill the container with water and repeat your observations. Make two more drawings.

3 Compare the observations of the stick in the water with the stick in the empty container. Note any differences.

4 Share your observations with the class and try to explain them.

a   b   c   d

Look at the four pictures of a stick in a jar with water. Which one shows the correct result? Tell the class what you think.

When light travels through transparent materials, such as air, water and glass, its speed changes as it moves from one material to another. This produces a 'bent' image as the path of the light is changed. This process is called refraction.

Lenses are made specially to use the refraction of light to focus images onto our eyes or onto a screen, for example. At the cinema, the film is projected from a machine with several lenses, called a projector. The small image on the film is magnified to fill the enormous screen, so that everyone in the cinema can see it at the same time.

People who have eyesight problems are helped by wearing glasses/spectacles, which have lenses. These bend the light as it enters the eyes and focus the images more precisely on the retina, which improves the vision of the person wearing them.

Activity 20

You will need: a source of sound.

1. Set up the sound source and form a circle around it. Observe what you can hear.

2. Slowly move back, further and further from the source of sound. Make a note of what happens as you get further and further from the source of sound.

3. Go as far away from the sound source as you can, noting what happens all the time. Return to your seat and share with the class what you have observed.

4. Try to explain your observations.

Activity 21

You will need: pairs of items like those in the pictures above, something to hit them with, musical instruments, water, paper and a pencil.

Pair of objects	Description of the sounds
1. Metal cup Plastic cup	
2. Empty bottle Half-filled bottle	
3.	
4.	
5.	

1. Draw a table for recording the results of your investigation of sounds travelling through the various materials you have collected.

2. Hit or blow each pair of objects and listen to the pitch, volume and other features of the sounds that are produced. Compare them before you make a note of your observations.

(3) When you have investigated all the sounds, look at your table of results and look for patterns.

(4) Discuss the results with your group.

(5) Share your conclusions with the class and try to explain them.

Both sound and light travel away from their sources in all directions. The further away we are from the source, the weaker the light or sound when it reaches our eyes or ears. If we are too far away, we will not be able to detect the light or the sound. This was demonstrated in the two activities when the class formed a circle around a source and then moved further and further away from it. Everyone in the circle could detect the light or sound; this showed that the light and sound travelled away from their sources in all directions.

The activities also revealed that it became harder and harder to detect the light and sound as the distance between them and your sense organs increased.

Sound travels through different materials – solids, liquids and gases. We live in air, so our ears are adapted to receive sounds transmitted through air. Animals such as whales and dolphins live in water and their ears are adapted to receive sounds transmitted through water. We all know that solids transmit sounds. Even walls and windows cannot stop sounds travelling in or out of buildings. Musical instruments are made of various solid materials and when they are hit, blown, plucked or bowed, they transmit sounds. The pitch, volume and quality of the sounds vary, depending on the size, shape and material of the instrument. Wood and metal are good sound transmitters and so many musical instruments are made of these solid materials.

Systems

Roots, shoots and flowers

In this unit, three of the systems of plants will be investigated. They are the root system, the shoot system and the reproductive system.

Roots

Activity 1

. You will need: a spade or garden fork, paper and a pencil.

1 Go outside with the garden tool and find some plants that you can dig up. Carefully dig up different types of plants. Be careful not to break off the roots because you will use them for making comparisons.

2 When you have found two very different root systems, take the plants back to class and examine them closely. Compare the form of the roots on each plant. Draw pictures of each one to illustrate the main differences and similarities. If you know the names of the plants, add them to your drawings.

maize

carrot

tap root   fibrous root

3 Look at the pictures of the two types of root systems and compare them with the plants you have drawn. Decide which system matches each of your plants. If you did not have an example of both types, check with other groups to see if they can let you have the missing one.

Root systems are of two basic types – the fibrous root system and the tap root system. The fibrous root system has many roots of equal size, all starting from the bottom of the stem. This is shown in the picture of the maize plant roots.

The tap root system has one main root and smaller roots growing out of it. This is shown in the picture of the carrot plant roots.

The basic structure of all roots is the same. They are branched and spread out into the soil in which they are growing.

Activity 2

You will need: a jar or other transparent container, water, a complete plant, a plastic bag, a wire or string tie for the bag, a ruler, a pencil and paper, graph paper.

1. Put water into the container to the top. Carefully dig up a small plant, taking care not to break off the roots.

2. Lower the plant into the container and allow the water to overflow. Wrap the plastic bag around the container and use the tie to close it tightly around the stem of the plant. This should hold the stem above the water level.

3. Draw a table for recording the water level each day over at least a week.

4. Measure the height of the water in the container and record the date and the height in the table.

5. Repeat the measurement and recording each day.

6. When the period of measuring is over, use the measurements you have recorded to make a graph showing what happened to the water level. The horizontal axis should show the days and the vertical axis should show the level of the water. Write a report of what you did in the investigation.

7. Look at the graph and use it to make a conclusion about the function of the roots.

8 Write your ideas down and then share them with the class. Discuss the conclusions that different pupils have.

9 Try to answer these questions:
Why was the top of the container covered with a plastic bag and closed around the stem?
Why did the level of the water change?
What would have happened if the roots had been cut off before the plant was put in the container?

bag closed
with a tie

a

b

Look at the two plants which the children were trying to pull out of the soil. Are they the same or different? Can you tell which weed was easy and which was hard to pull up? How can you tell? Share your answers with the class.

Copy and complete these sentences.

Here are the words you will need: *branches*, *lot*, *roots*, *hard*, *easy*, *holding*, *soil*, *pull*.

(a) It was _ _ _ _ to _ _ _ _ up most weeds because their _ _ _ _ _ did not have many _ _ _ _ _ _ _ _ .

(b) The plant that was _ _ _ _ to pull up was _ _ _ _ _ _ _ _ onto a _ _ _ of _ _ _ _ .

The root system has two functions. It collects water and minerals from the soil and transports them up into the stem of the plant. Without this supply from the roots, the plant would die.

The root system also provides the plant with anchorage. Systems with many and deep branches hold on to the soil best. They prevent the plant from being blown down, or uprooted by floods and heavy rain.

Activity 3

1. Set up the jars with the seeds, water and absorbent material so that the seeds will germinate.

2. Put the jars in a safe place and look at them each day, making sure that they do not dry out.

3. When a seed germinates, measure the length and spread of the roots. Record them in a table, together with the date each time you make a measurement. Draw a picture of the roots each day to show the shape of the system.

4. Continue with the observations, measurements, drawings and recording for as long as you can.

You will need: large seeds such as bean or maize, transparent containers in which to germinate the seeds, water, absorbent paper or cotton wool, a ruler, paper and a pencil, books and other resources about root systems.

5. Use the recorded measurements and drawings to describe what happens to the root system as it grows and develops. Try to explain why these changes happen. Discuss your ideas with your group.

6. Use the books and other resources to find out what they tell you about the way roots change.

7. Make notes of what you find and share your conclusions with the class.

As a plant grows, the number of stems and leaves increase. This means that more and more water is needed to keep them alive. The roots need to grow and keep pace with the parts of the plant above the ground. In this way they can increase the amount of water and minerals that they collect from the soil. More and more branches appear in the root system and they spread further, drawing on a larger volume of soil.

The other change as the plant grows is that its stems spread wider and become taller. This means that it is easier and easier for it to fall over. To reduce the chances of this, the roots also need to spread wider and deeper to provide enough anchorage for the plant.

Shoots

Activity 4

You will need: books and other resources about shoots/stems, paper and a pencil.

1 Discuss with the class and your teacher what you could investigate about plant shoot systems.

2 Choose your group's area of enquiry and use the resources to collect information. Make notes and drawings to record your findings.

3 Present the group's findings to the class. Listen to what other groups have to tell you and ask them questions about their area of enquiry.

4 Display the group's work on shoots.

Activity 5

You will need: a container for water, a piece of a soft-stemmed plant, coloured water, a knife or scissors, paper and a pencil, a magnifying glass/hand lens.

1 Put the coloured water in the container and stand the shoots in it. Put the container in a safe place.

2 Observe the shoots each day and record what you see, in drawings and notes.

3 Keep the observations and records going for as long as possible. When they end, use the knife or scissors to cut through the shoot near the end that was standing in the water. Use the magnifying glass to see more clearly what the inside of the shoot looks like. Draw what you see.

4 Cut the shoot in other places and draw what you see each time.

5 Cut a piece of the shoot along its length and carefully observe what the inside looks like. Draw what you see.

6 Use your drawings and notes to produce a full report of what happened, in the correct sequence. Try to explain what you have observed. Display your report and look at those produced by others.

Shoot systems, like root systems, vary in their form. Some are permanent (perennial) and each year new branches develop and the whole system grows longer and thicker. Trees and shrubs have this kind of woody shoot system.

Other systems are annual. They grow to their full size and then die within one year. Many weeds, garden flowers and crops have shoot systems of this kind, e.g. maize, sunflowers. Their shoots may be thin and soft, or more thick and woody. They may have many branches, few or none.

The shoot system has several functions. It carries the leaves and holds them up in the sunlight. This allows the leaves to carry out their function of making food, by the process of photosynthesis.

The flowers also develop on the shoots and these in turn develop into the fruits. Shoots carry the flowers into positions where they can perform their own function, which is reproduction.

The shoot system also connects the leaves, flowers and fruits to one another and to the root system. This is vital for the transportation of water and minerals from the roots to all parts of the plant and for the distribution of food from the leaves to all the other parts of the plant, which are unable to feed themselves. This means that the shoots are a busy 'highway', with movement both up and down of water with various materials dissolved in it.

water and minerals from the roots

food from the leaves

Flowers

Flowers are the reproductive organs of flowering plants. Their function is the reproduction of the plant.

Sepals are the outside parts, which are usually green. They are wrapped round the flower when it is in the bud.

Petals are usually the largest and most colourful parts of the flower. They are just inside the sepals and they unfold when the bud opens.

Activity 6

You will need: paper and a pencil, a ruler, hand lens, a bag for collecting flowers, books and other resources about flowers.

1 Draw a table like the one below for recording your observations of flowers.

Flower name	Colour	Shape	Smell	Number of petals	Number of sepals	Size of petals

Make sure that the 'shape' column is big enough for you to draw a small picture of each flower.

2 Take the bag with you when you go out on the field trip to find and collect different flowers. Collect only <u>one</u> of each kind. Remember to collect wild flowers/weeds, as well as flowers from trees, shrubs and garden plants. DO NOT pick flowers from gardens belonging to other people. Use your own, or the school garden.

3 Back in class, observe eack flower and record the details in the table.

4 If you do not know the names of the flowers use books and other resources to identify them.

5 Give a report to the class of what you have found out about flowers.

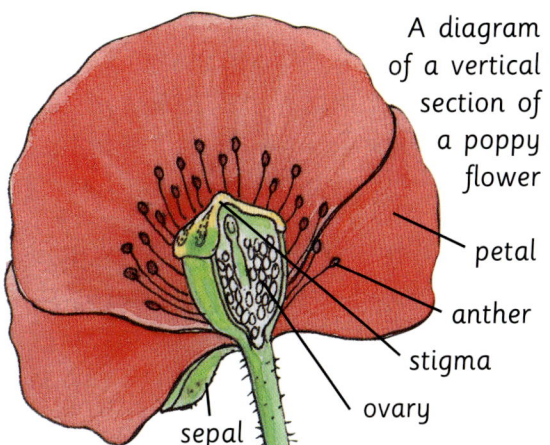

A diagram of a vertical section of a poppy flower

petal

anther

stigma

ovary

sepal

Activity 7

1 Copy the diagram of the flower above, including the labels of the parts.

You will need: paper and a pencil, a ruler, books and other resources about flowers.

2 Draw a table with one column for 'parts of the flower' and another for 'functions'.

3 Use the resources to collect information about the functions of the parts labelled in the diagram. Enter the information in the table.

4 Use the information to colour the female reproductive organs one colour, the male reproductive organs another colour, the sepals green and the petals a fourth colour.

Reproduction in flowers is sexual. That means that two different cells have to combine to produce a new plant. The female reproductive organs produce the female sex cells, called ovules. These are made in the ovary, which is at the bottom of the flower. The ovules are attached to the walls of the ovary, so they are not able to travel in search of the male sex cells.

The male sex cells are called pollen and they are made in the anthers, which are the male reproductive organs. The anthers split and release the pollen grains so that they are free to travel to the ovules.

Those flowers that have petals and scent use these features to help them bring the pollen to the stigmas, which are on top of the ovary. The colours of the petals and the scent of the flowers attract certain insects that feed on pollen and nectar, a sugary liquid made at the base of some flowers. As the insects push into the flowers, they bump against the anthers and showers of pollen grains fall out. Some fall onto the sticky surfaces of the stigmas and get stuck there.

Other pollen grains get caught in hairs on the insects' bodies. These grains can also get stuck to the stigmas. Insects fly from flower to flower, picking up and dropping pollen as they go.

When the pollen lands on the stigma, the flower has been pollinated. This process is called insect pollination.

Some flowers do not depend on insects to be pollinated. The grasses, like maize, rice and wheat, for example, do not have colourful petals or any other feature to attract insects to them. Their anthers hang out of the flowers and, as the pollen falls out of them, the wind blows it away and some of it lands on the sticky stigmas of other plants. This process is called wind pollination.

Once the pollen has been released from the anthers, their function is finished and they die. The pollen grains on the stigmas grow tubes down towards the ovules. The male sex cells travel down the tubes and fertilise the ovules. Once this process of fertilisation is complete, the stigmas' work is done and they die.

Inside the ovary, the fertilised ovules begin to grow and develop into seeds. The wall of the ovary often changes too as this process goes on. In some plants the ovary becomes a large, fleshy fruit, with the seeds deep inside, e.g. orange, melon. Others, such as the coconut, are large but not fleshy, containing one seed inside each hard nut.

An apple

A melon

A coconut

Activity 8

You will need: a flower worksheet, knife, flower, a pencil, a hand lens, glue.

1. Use the knife to carefully cut the flower into its separate parts. Keep the parts in sets so that you can identify them.

2. Use the diagram on the worksheet to name each of the parts, using the hand lens to observe the parts closely.

3. Take one of each part and glue it on the worksheet opposite the diagram. Link each part with its name on the sheet by drawing a line between the two.

4. When it is complete, put it into your folder.

Human body systems

The human body has many systems, each one with its own function. Only three systems will be dealt with in this sub-unit: the **skeletal/muscular system,** the **excretory** system and the **reproductive** system. The organs and their functions in each system will be explored.

The skeletal/muscular system

Activity 9

1 Find a space where you can stand and move your limbs and body without bumping other people.

You will need: books and other resources about muscles, bones and joints, movement and locomotion, paper and a pencil.

2 Use each of your joints, one by one, to explore the full range of movements that you can make. Bend, stretch, turn, expand and contract the various parts of your body by using the flexibility given to you by your joints.

3 Sit down with your group and discuss what the bones, joints and muscles do when you move parts of your body. Point to examples of each part.

4 Use the resources to collect detailed information about the way the body moves its parts and how it moves from place to place (locomotion). Make notes and drawings of what you find out.

5 Display your work for the class to see and explain what it means. Look at what others have done, listen to them and ask them questions.

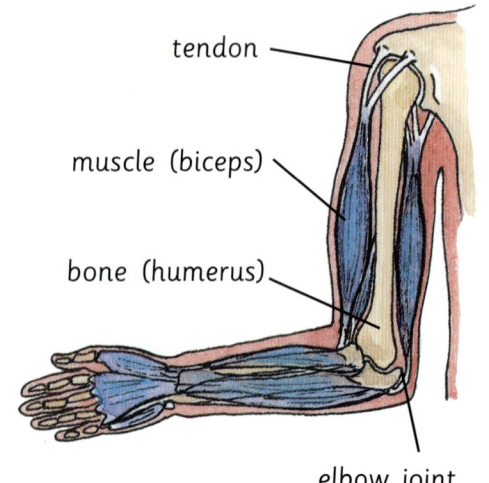

tendon

muscle (biceps)

bone (humerus)

elbow joint

The skeleton and the muscles attached to it are essential for movement and for locomotion. The skeleton is made up of 206 bones. They are hard and strong and, where two bones meet in a joint, there is a slippery surface on the end of each bone. This allows the bones to move smoothly over one another in the joint. The bones in the joint are held together by ligaments.

Bones and joints cannot produce movement on their own. It is the muscles that move the joints and this leads to movements and locomotion.

The muscles are attached to the bones by tendons. When the muscle contracts, it pulls on the tendon and this pulls on the bone. The result is some movement. This is what happens every time we walk, pick up a pencil, kick a ball, swim or eat, for example. The organs of the skeletal/muscular system – the bones, joints and muscles – have to work together to produce movement. Muscles without bones would not be able to support our body, or move it from place to place. A rigid skeleton without joints would not allow any movement either.

The excretory system

The body is kept alive by many processes and they make waste products. If these wastes are not removed from the body, they can damage it. The process of removing waste products from the body is called excretion. There are several excretory organs, each one dealing with particular types of waste.

Activity 10

1. Think of the things that you have to do every day to stay alive and well. Share your ideas with your group.

   You will need: paper and a pencil.

2. Sort out those which are to do with getting rid of waste products from the body. Discuss with your group which organs are used to get rid of each type of waste.

3. Sort out the jumbled names of organs and waste products in the lists below.

   Organs: nksi siynked glnsu ielrv

   Wastes: elbi stawe niuer baocrn dodieix

4 Match the organs and the wastes and write down the pairs.

5 Share your answers with the class.

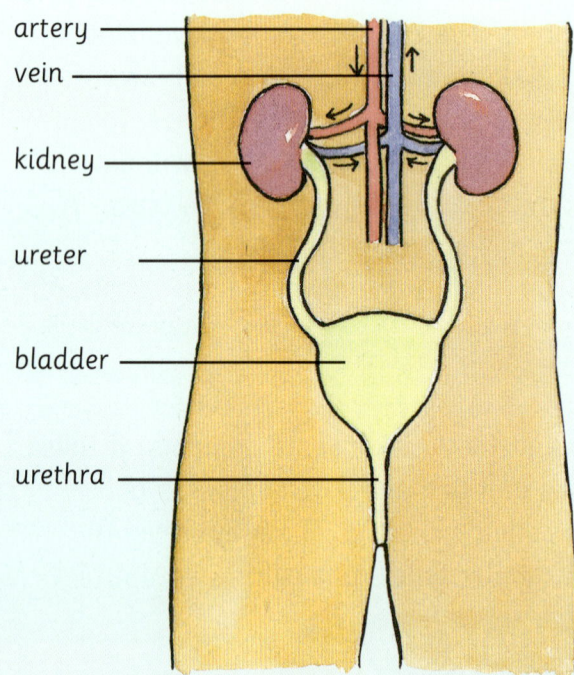

artery

vein

kidney

ureter

bladder

urethra

This diagram shows one part of the excretory system of the human body. It is sometimes called the urinary system, because its function is to get rid of urine. This is mostly water, with several materials dissolved in it. The main waste product is urea, a poisonous material produced by the liver from other waste materials carried in the blood. (The liver also produces bile, a waste product which it passes as a liquid into the intestine.)

The blood carries the urea and other poisonous materials to the pair of organs that can remove them and clean the blood – the kidneys. They are located on either side, at the level of our lowest ribs. In adults they are about 12 cm long.

The blood carrying the waste products enters the kidneys in the arteries. It has all the wastes taken out, along with a lot of water. Then the cleaned blood leaves the kidneys in the veins and returns to the heart, to be sent out eventually to all parts of the body.

The watery urine leaves the kidneys in tubes called the ureters, which carry it down to an elastic bag of muscle, called the bladder. The urine is stored there until the person feels that they need to empty the bladder by urinating. Babies do not have control of their bladders, but as we grow and develop we learn how to keep the exit from the bladder closed and so we can choose when to urinate.

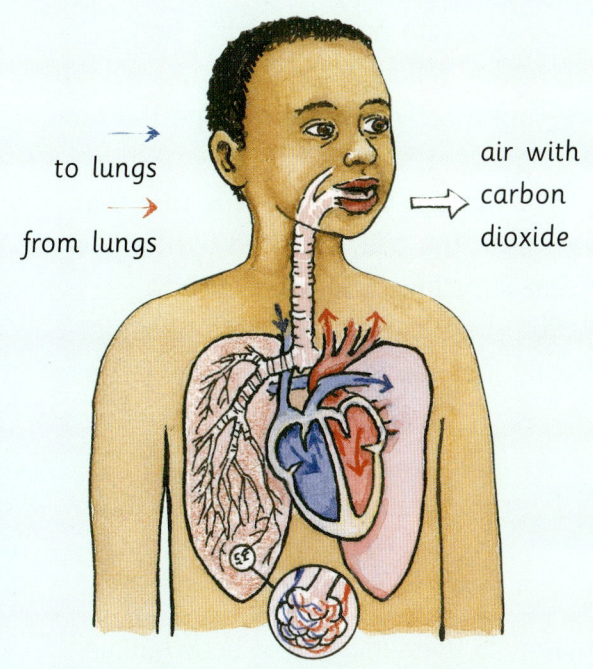

to lungs

from lungs

air with carbon dioxide

This diagram shows a second part of the excretory system of the human body. It is sometimes called the respiratory system. The two lungs are the excretory organs which get rid of the gas called carbon dioxide. This gas is a waste product from all the cells of the body. It dissolves in the blood and is carried away from all parts of the body to the lungs. The lungs are full of tiny sacs filled with the air that has been breathed in. These tiny sacs are surrounded by tiny blood vessels. The process of gaseous exchange takes place in these tiny sacs. Oxygen dissolves into the blood and is carried away to all the cells of the body. At the same time, carbon dioxide leaves the blood and is released into the air, filling the tiny sacs. So there is an exchange of gases – oxygen *in* and carbon dioxide *out*. When we breathe out, we excrete the carbon dioxide into the atmosphere, through our nose and mouth.

The skin is the third organ in the excretory system. It shares with the kidneys the removal of some wastes from the blood, especially urea and some salts. The surface of the skin is covered with tiny holes, called pores. At the bottom of the tubes leading up to the holes are special bodies called 'sweat glands'. They excrete sweat, a mixture of water, urea and other waste materials.

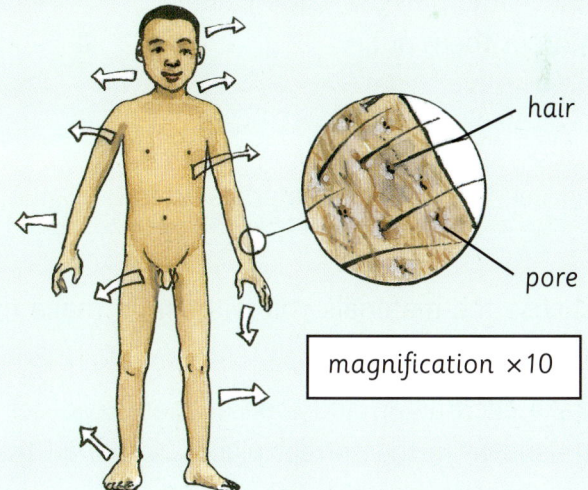

hair

pore

magnification ×10

We produce sweat all the time, but we only notice it when we have been very active, or the weather is very hot. Then the tiny drops of sweat join together on our skin and we see them and feel them running over it. This helps to cool us down at the same time as excreting wastes from our bodies.

In everyday speech, we often refer to faeces as a 'waste product' of the body. Scientists do not usually include faeces in excretion, because they are not really a *product* of the body's living processes. Faeces are mostly made of undigested food, which has not been *made* by the body; it was put into the body through the mouth and it passes out again from the rectum, through the anus. It is true that the brown colour of faeces is due to materials from the bile, excreted by the liver and added to the food as it passed through the intestine. Scientists refer to defecation as 'elimination', rather than excretion, because the products have not been removed from the blood: faeces are mostly just solid bits and pieces left over from digestion.

Activity 11

Either:

You will need: paper and a pencil, colouring materials, scissors, glue, a body outline.

1 Use the diagrams of the excretory organs to draw copies of each one.

2 Colour them all the same colour and cut them out.

3 Glue each organ on the body outline in the correct place. Label all the organs.

4 Display your finished diagram

Or:

You will need: modelling materials, scissors, glue.

1 Choose the materials you will use to make models of the excretory organs.

2 Make each organ and then fit them together inside a body shape so that they are shown in the correct places. Label all the organs.

3 Display your model and look at what other groups have produced.

Reproductive system

Reproduction is one of the characteristics of all living things. In this way, new individual plants and animals are produced, to replace those that die. Death is a fact of life for all living things, including human beings. Reproduction is our way of keeping the human race alive.

In human beings, reproduction is sexual. This means that it involves two different cells combining to produce a baby human. Not all living things reproduce in a sexual way. Some simple living things just split themselves in two and the two halves then live as new individuals!

We cannot do this.

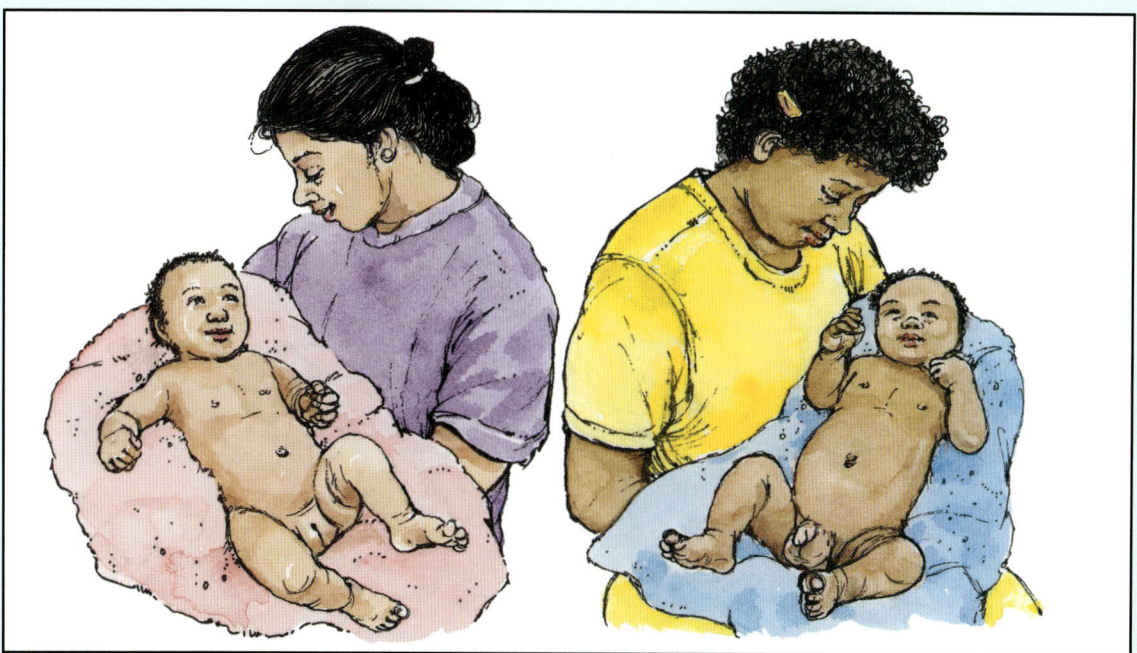

There are two types of human beings – girls and boys, who grow up to be women and men. Just as the flowers have different organs to produce the two types of cells for reproduction (the pollen and the ovules), so people either have female organs or male organs. Female reproductive organs are found in one sex – the girls and women. Male reproductive organs are found in the other sex – the boys and men. This is why we have different names for the two sexes and this is how we know which sex we belong to. When a baby is born, it is the reproductive organs that identify it as a girl or a boy.

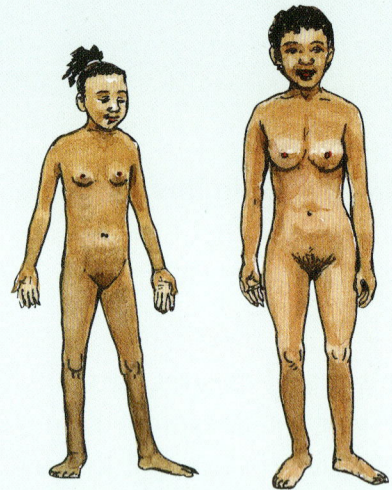
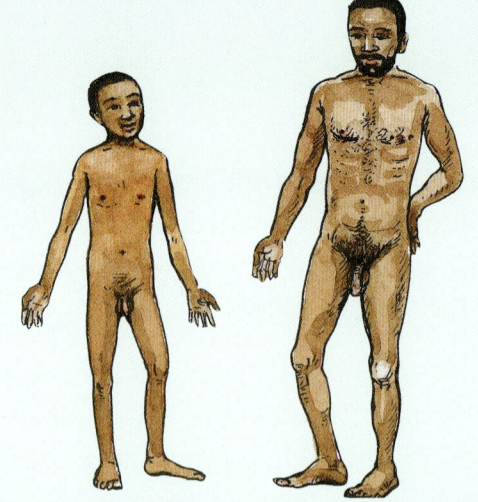

The pictures show children like you and two adults. Look for similarities and differences between the four people in the pictures. Tell your group what you see and discuss the reasons for the differences and the similarities.

Each sex makes one kind of sex cell. The female sex cells are called eggs. The male sex cells are called sperm. Women and men have special organs that produce these cells. They also have organs that make it possible for these two types of cell to be brought together, so that a baby can be produced.

Eggs are made in the two ovaries, which are inside the female body. Each ovary in an adult woman is about 3.5 cm long. The eggs are normally released from the ovaries one at a time, each month. This process of releasing eggs does not begin until a girl reaches puberty. This is the time when the girl starts the process of changing from a child into an adult. This 'in-between' stage of life is called adolescence. Puberty, when the first eggs are released, usually happens between the ages of 11 and 13.

a

uterus

vagina

vulva

ovary

b

uterus

ovary

vagina

vulva

bladder

Female reproductive organs
a) from the front
b) from the side

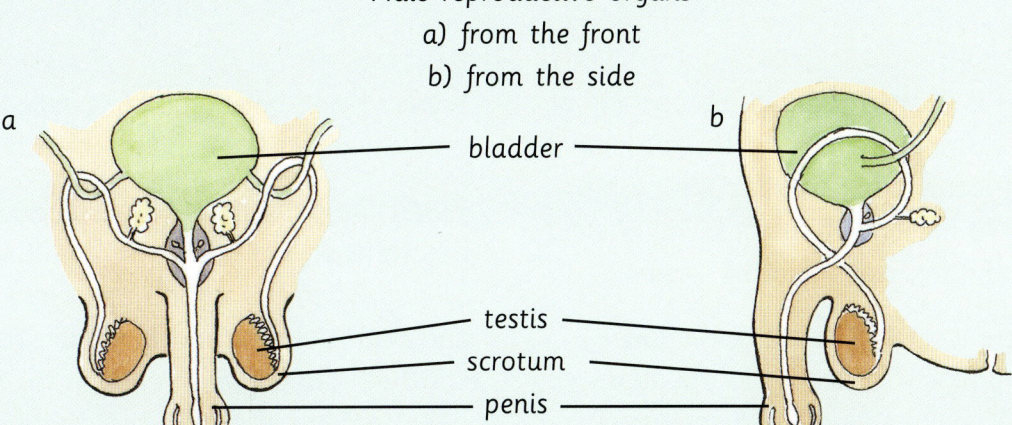

Male reproductive organs
a) from the front
b) from the side

a

b

bladder

testis

scrotum

penis

Sperm are made in the two testes, which are inside a bag of skin between the legs, called the scrotum. Each testis in an adult man is about 6 cm long. Sperm are made in very large numbers all the time, from puberty onwards (tens of millions each day). As with girls, puberty in boys marks the beginning of adolescence, when the boy starts the process of changing into an adult. Puberty in boys, when the first sperm are made, usually happens between the ages of 13 and 15.

Most types of animals lay eggs, from which the young hatch later. In mammals, the baby develops inside the body of the mother. The uterus, or womb, is the organ in which baby mammals grow and develop. This organ is completely inside the female body, surrounded and protected by the muscles of the abdomen/belly at the front and the spine and body wall at the back.

This means that there has to be some way of bringing the sperm and egg together inside the female, so that they can combine and produce a baby. It also means that there must be some way for the fully developed baby to leave the uterus and come out into the world, at birth.

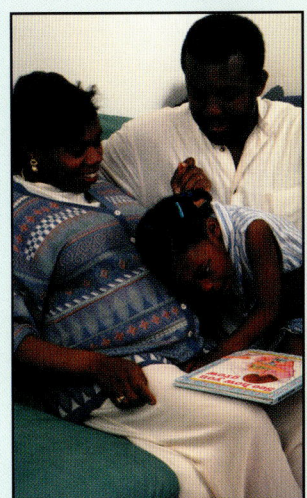

The female reproductive organs are designed to allow both these things to happen. The vulva is the opening between the legs which is both an entrance and an exit.

The vulva is the entrance for the penis. This part of the male reproductive system is used to deliver the sperm deep inside the female body, so that they are in a good position to meet the egg and fertilise it. During sexual intercourse, the penis becomes erect and it is pushed through the opening of the vulva into the vagina. The movement of the penis during sexual intercourse leads to millions of sperm being pushed out from the penis into the vagina.

If there is an egg in one of the tubes leading from the ovaries to the uterus, then it may be found by a sperm cell and fertilised. If this happens, the egg then settles in the uterus and begins its growth and development into a new

human being. This process is called pregnancy and lasts about nine months.

When the baby is fully developed it has to leave the body of its mother. This is when the vagina and vulva carry out their second function. They can both stretch wide enough for the baby to come out from the mother's body. It is delivered through the vulva and begins its independent life as a new human being.

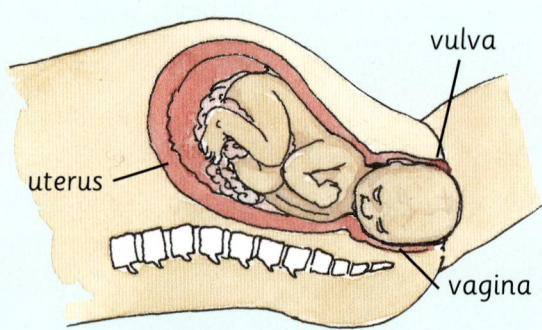

Activity 12

1 Discuss with your group which system you will investigate using the resources available.

You will need: books and other resources with information about excretion and reproduction, paper and a pencil.

2 Share out the work of collecting detailed information about your chosen system.

3 Make notes and drawings as records of what you find out.

4 Produce a display of your findings to share with the class. Answer any questions that other pupils may have.

5 Look at the displays of work from other groups and ask them questions about what you see.

The environment and us

Communicable diseases

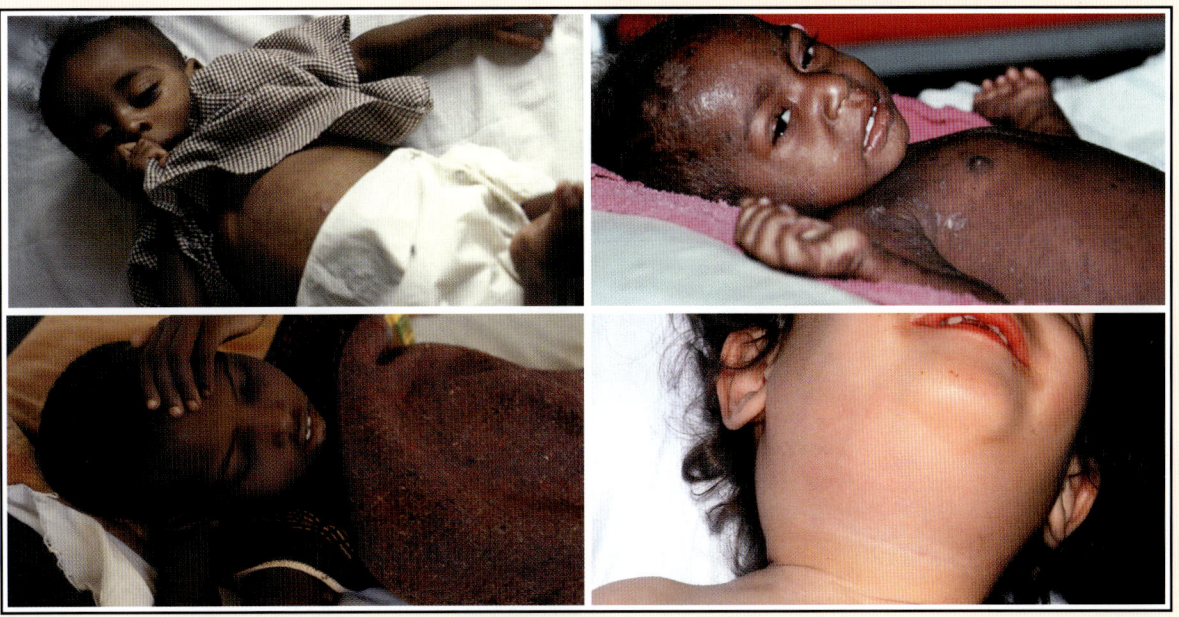

Activity 1

You will need: paper and a pencil.

1. Share with your group any experiences you have had of communicable diseases.

2. Give your reasons for thinking that the diseases were communicable.

③ Write down all the diseases mentioned by your group, with the reasons why people think they are communicable.

④ Share your list with the class and compare it with those from other groups.

A deficiency disease, such as scurvy or kwashiorkor, is not communicable. It is not caused by any kind of infection. It is the lack of some nutrient which causes it.

A disease such as diabetes or asthma is also not communicable. It is caused by some fault in the way the body is functioning, rather than by an invasion of the body by another living organism.

Flu, the common cold, conjunctivitis, cholera, AIDS, tuberculosis (TB), dysentery, measles, pneumonia, malaria and chicken pox are all communicable diseases. These diseases are all caused by living organisms, which have somehow got into the body. These organisms are spread and passed on from one infected person to another: they are communicated.

Activity 2

You will need: paper and a pencil, books and other resources about diseases.

① Discuss with your group the various ways in which diseases are spread from one person to another.

② Make a list of all the methods that the group knows.

③ Try to find more methods from the books and other resources. Find diseases that are examples of each method of transmission. List them under the methods.

④ Share your lists with the class and attend to what other groups tell you. Add any methods or diseases that are missing from your lists.

The pictures show the different ways in which diseases can be communicated/transmitted. Match the methods with the names of the diseases listed below. Write down the letters on the pictures and the names of the diseases.

measles   TB   flu   AIDS   dysentery   malaria   food poisoning

Our bodies are open to the outside world. We have many holes through which we can be invaded by other organisms, which can then live in our bodies.

We cannot stop breathing: we have to take in air all the time and this gives some organisms the chance to be carried into our mouths, noses and lungs. Flu, the common cold and TB are examples of diseases that use this route into our bodies, in contaminated air.

We cannot stop eating: we have to take food into our bodies every day and this gives other organisms their chance to get inside us. They are carried in the food. Some types of worms, plus the germs that cause food poisoning and dysentery can be passed on in this way. Flies can carry the germs from place to place, landing on faeces from infected people and then on our food.

We cannot stop drinking: we have to take in water several times a day and this is a perfect way for some diseases to spread. If the water supply becomes contaminated with the organisms that cause the disease, then everyone who uses the water will swallow the organisms and be infected. Diseases such as cholera, polio and typhoid are communicated this way.

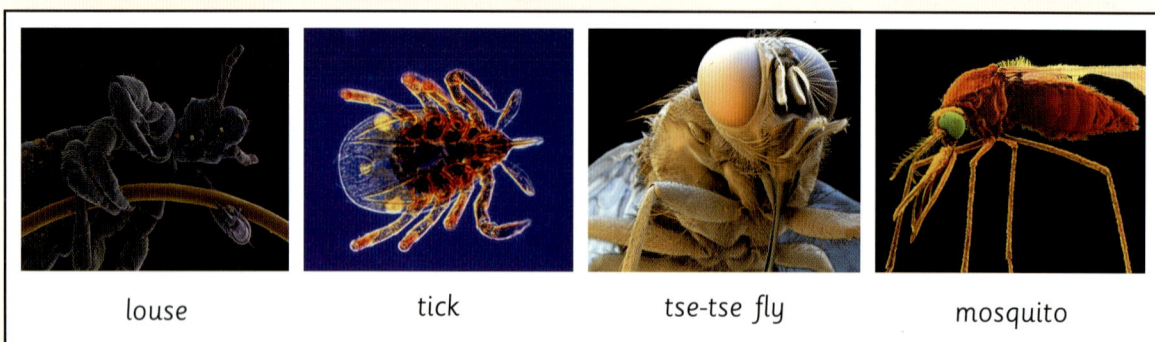

louse   tick   tse-tse fly   mosquito

We cannot keep our skin completely covered the whole time: we have to wear clothes that suit the weather. This means that some parts of our bodies are left uncovered and certain insects can bite us. Some of them carry the organisms that cause diseases. The mosquito can transmit malaria. The tsetse fly can transmit sleeping sickness. Lice and ticks can transmit various fevers.

During sexual intercourse, the bodies of the man and the woman are joined together. This allows germs to pass from one to the other, in either direction. There are several diseases that are spread this way. The most serious is AIDS, for which there is no cure at present. Diseases spread this way are called STDs – sexually transmitted diseases. If neither partner is infected with a disease, then nothing can be transmitted.

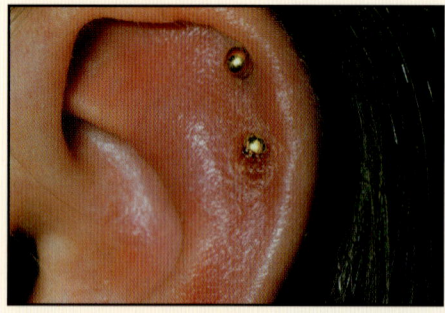

Needles used to inject drugs and instruments used to pierce the earlobes or other parts of the body for jewellery to be fitted can also carry diseases into our bodies. That is why these things must be clean before anyone uses them. Making holes in our skin is always a dangerous thing to do and it is best to avoid it as much as possible. Even doctors in hospitals must take great care not to let wounds, made during operations, become infected. Blood poisoning can develop and it can kill the patient if it is not quickly treated.

Sharing beds, towels, clothes and eating or drinking together can spread some diseases that are highly infectious. Sometimes it is necessary to isolate people who are infected so that they do not spread the disease.

One very dangerous disease is transmitted when a person is bitten by an infected dog, cat or bat. This disease is rabies. If the infected person is not treated very quickly after being bitten, they will die within a short time.

Activity 3

You will need: paper and a pencil, books and other resources about diseases.

1 Use the resources to research either epidemics or pandemics.

2 Look for explanations of why an outbreak of a disease becomes an epidemic or a pandemic. Make a list of the factors that lead to this result.

3 List as many examples of disease epidemics or pandemics as you can find, with dates and the areas that were affected.

4 Share your findings with the class in some way, e.g. you can produce a chart, notes with drawings and maps, or be interviewed by other groups.

Diseases can only spread if there is some way for them to pass from one person to another. If the air, food and water supplies are kept clean and free of infections, then the chances of diseases spreading are very much reduced. These are the most common routes for widespread infection.

During catastrophes, such as floods, earthquakes, wars and famines, the normal controls on diseases break down. Many people may end up living together in bad conditions, where there is not enough clean water or proper drainage for toilets and crowds sleep and live close to one another. Such conditions are perfect for an outbreak of disease to spread quickly. Flies, lice and other biting insects can carry germs from one person to another. Faeces get mixed with the water supply. Food is not kept clean, or covered, or cool. It quickly 'goes bad' and makes people ill.

One infected person in a family cannot be kept apart from all the others, so their infection gets passed on. This is how many epidemics start, in situations where control of diseases has broken down. Poverty is also often a factor in epidemics, as the people cannot move away to cleaner, less crowded places. Nor can they buy the drugs to treat the diseases, so they remain infected and act as carriers within their communities.

Pandemics need some added factor that can carry the infection over great distances, so that the outbreak appears in many parts of the world. In the past, many traders carried their goods by ship and infections had to survive the long sea journeys somehow. Plague, for example, was brought to Europe from Asia, carried by fleas which lived on the rats living on the ships. The disease killed millions in Europe in the 14th century. In modern times, the aircraft has made it much easier for infections to move from one side of the world to another. Infected individuals can travel far from the place where they became infected and carry diseases to places where everyone is free of them. This gives the organisms many new opportunities to find people in which to live.

The misuse of drugs

A drug is any substance, other than food, which causes changes in the body when it is swallowed, breathed in, injected or applied to the body in some other way. There are three sets of drugs: • prescription drugs
   • over-the-counter drugs
   • prohibited drugs.

Activity 4

1 Discuss the three different types of drugs. Share your ideas about why some drugs are in one set and not in another.

You will need: paper and a pencil.

3 Share your lists and ideas with the class.

2 Write down examples of each set of drugs.

4 Add to your lists examples given by other groups.

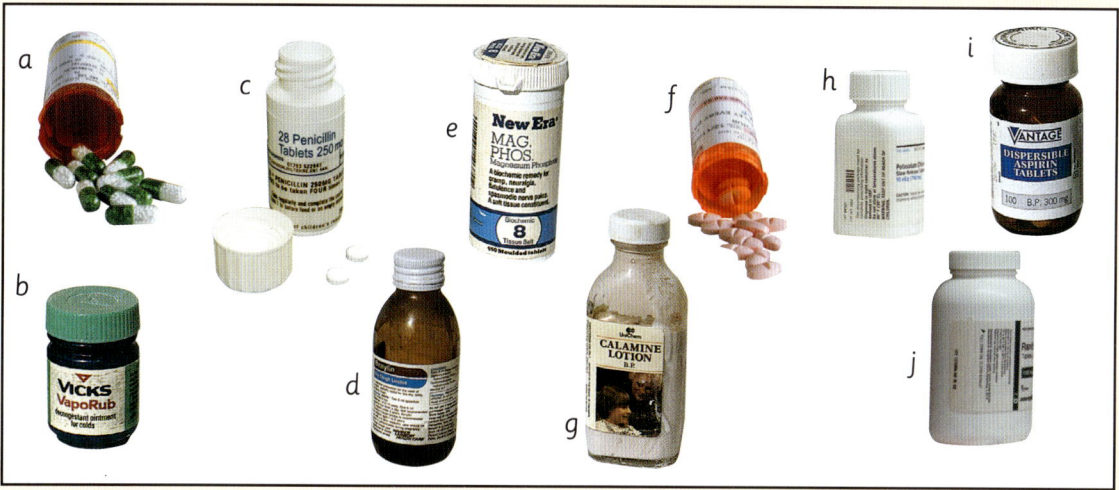

Look at the pictures of various drugs and sort them out into two sets:

- prescription drugs
- over-the-counter drugs.

Write down the two sets of letters which label the drugs. Share your sets with the class.

Activity 5

You will need: empty drug packets and bottles, paper and a pencil.

1. Draw a table for recording information about the drugs that were in the packets and bottles.

Drug	Use	Dosage	Expiry date	Side effects	Warnings
A					
B					
C					

2. Copy the information from the packets and bottles, recording it in your table.

3 Decide how you will present the information you have collected. Sort the drugs out in some way, e.g. sets for each type of drug use, sets for each type of warning.

4 Display the drug containers.

5 Share your findings with the class and discuss what you have all found out about the proper use of drugs.

Drugs bought in pharmacies and drug stores all have very important information on their packaging. This is given to protect people from harm. Drugs are dangerous. This is true of prescription and over-the-counter drugs, as well as the prohibited drugs.

To use any drug safely, we must follow the instructions about the dose, the age of the patient, the warnings and side effects. More of a drug does not mean more benefit! It might mean death or serious damage to our bodies. For example, organs such as the liver and the kidneys can be damaged by high drug doses and it may not be possible for them to recover.

Children are in the greatest danger from drugs, because their bodies are smaller and so they are more easily damaged by an overdose of drugs. This is why drugs must be stored safely at home, in a place where young children cannot get them.

Activity 6

1 Discuss with your group:

'What are the benefits of drugs?'

'What are the harmful effects of drugs?'

2 Keep notes of the group's answers to these two questions.

3 Share the group's answers with the class.

Prescription and over-the-counter drugs have many benefits. Some can cure diseases, killing the organisms that have invaded our bodies. Others can remove the symptoms, such as the sneezing and coughing of a flu infection, without killing the virus causing the flu. Some can prevent us being infected. They give us protection against the disease. Anti-malarial drugs are of this kind. If our body is not working properly, causing a disease such as diabetes, drugs can sometimes correct the fault and allow the person to go on living.

Some people misuse prescription drugs. They do not use them because they are sick, but because they like the effects that the drugs produce. Their feelings are affected and they can become addicted to taking the drugs. Such drug abuse is sad and dangerous. It means that the person cannot live a normal life and they are 'hooked' on the drug, unable to escape from it. Some people become criminals because of their addiction, stealing so that they can pay for the drugs they depend on for feeling 'good'. Addiction can also kill. Sometimes the drug kills directly, though often it causes accidents when the addict is not in complete control of what they are doing.

The two most common over-the counter-drugs are alcohol and tobacco, which are both addictive and legally available to adults. Like all other drugs, abuse can lead to serious damage to the body and even death.

Prohibited drugs are substances that the government has decided are not needed for treating disease and are too dangerous to be sold. Such drugs are often very addictive. Those which are injected, such as heroin, also carry the risk of infection with AIDS and other diseases, if addicts share needles. It is important to say NO! to such drug abuse for all these reasons, which are explained above:

- Drugs are dangerous. They can damage or kill the body.

- Drugs can be addictive, taking control of the person's life.

- Prohibited/illegal drugs are the cause of many crimes.

- Drug taking can carry infections into the user's body.

Activity 7

1 Use the books and other resources to do research on the use and misuse of drugs.

You will need: books and other resources about medicines and other drugs, paper and a pencil.

2 Use the following headings for your notes:

(i) the effects of the drug on the body

(ii) attitudes and behaviour that reduce drug misuse.

3 Choose a way of reporting your research to the class, e.g. a chart, a verbal report or a display.

Caring for the environment

People are part of the natural world. We are living things which depend on the natural resources of the planet for our lives – the air, the water, the soil, the plants and the animals. We cannot survive as individuals, or as a species, if we destroy our environment. It is foolish to behave as though our actions have no consequences. Some of the things we do are beneficial to the environment. Others are neutral – they do no harm or good.

The danger to the environment comes from those things that damage the natural world. In the past, people did not know or understand 'the web' of life and how we are linked to the living and non-living

Activity 8

You will need: books and other resources with information about the environment, paper and a pencil.

1   Discuss with your group what 'environment' means. Keep notes of what the group thinks.

2   Decide who will do research on particular aspects of the environment:
Choose the local, national or global level.
Choose which feature of the environment you will research.

Focus on the sustainable development of that feature, e.g. forest, water.

3   Collect information from the books and other resources about your chosen environmental feature.
Make notes and draw pictures, maps and graphs to illustrate your findings.

4   Put all your work in a portfolio so that the class can share your findings.

parts of the environment. Our ancestors cultivated the land, dug up minerals, made wastes and threw them to one side without much thought about the long-term results of such activities. Scientists and others have gradually built up a better understanding of how selfish, stupid and dangerous it is for people to live in such a careless way. People all around the world are now aware of their individual responsibility to care for the environment. This unit is included in your studies because you also need to be aware and learn what you can do – now, while you are still young – to share in the protection of your planet.

Activity 9

You will need: books and other resources about environmental damage, paper and a pencil.

1. Use the resources to do research on one of the following environmental problems:
   - Pesticides used by farmers
   - Industrial waste pollution
   - Deforestation
   - Endangered species
   - Acid rain
   - CFCs and the ozone layer
   - The greenhouse effect and global warming
   - Smog
   - Misuse of water resources
   - Noise pollution
   - Domestic waste disposal

2 Explore the harmful effects of your chosen problem at the local, national and global levels. Try to find examples of the damage done at all three levels.

3 Try to find out what can be done to prevent the damage.

4 Try to find out what can be done to solve the problems we have created.

5 Make notes and drawings of what you find.

6 Choose how your group will present your report to the class and prepare your presentation.

7 Attend to the presentations of other groups and record the information that they present to you. Ask them questions about anything that interests you or that is not clear.

Pollution is one very big issue that affects the environment at the local level, as well as globally. We cannot completely stop producing various types of waste – from our bodies, from our factories, from our burning of fuels, from our domestic life. The challenge for everyone is to reduce the amount whenever possible, then to re-use items and materials whenever we can, and finally to recycle waste as much as possible.

Careless waste disposal leads to pollution of the basic components of the environment – the air, the water and the soil. So it not only spoils the place where it is dumped, its effects can spread far from its source.

The trees in this forest have been killed by acid rain. The air was polluted by smoke in countries hundreds of kilometres away. The wind blew the smoke towards the forest and, when it rained, the pollution in the smoke turned the rain water into acid. This killed the trees.

This nuclear power station at Chernobyl, in the Ukraine, exploded in 1986, releasing large amounts of toxic radioactive material into the air. Swedish scientists detected this pollution in the air as the winds blew it towards them from the Ukraine, many kilometres away. Soil, water and the air became polluted with the radioactive material. Many crops and animals in

several countries had to be destroyed, because they were too dangerous to eat. Thousands of people had to be moved from their homes because the environment around the power station was too dangerous. People have died and more will die in the future, because of this pollution.

These two types of bird are extinct. People hunted them until there were none left. Of course, they did not mean to kill them all. It was done accidentally, but

Dodo   Passenger Pigeon

there is no way of putting this mistake right. Now we are more aware of what we are doing. We know that there are many plants and animals that are in danger of extinction. We know that we will have to protect them, or they will also disappear completely.

The natural world survives by means of a number of 'cycles'. These depend on various processes linked together. Changes in one part of the cycle have effects in other parts too.

Human behaviour is a part of these cycles and we now realise that we can do damage to them.

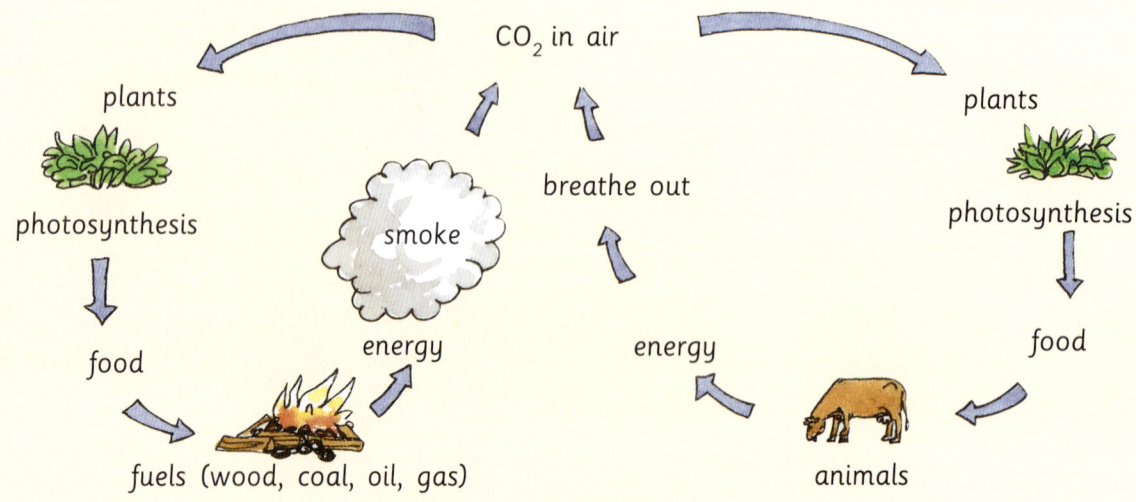

The Carbon Cycle

Look at the diagram of the carbon cycle above and explain how it is affected by some environmental problems. Tell your group what you think.

The Water Cycle

Look at the diagram of the water cycle above and explain how it is affected by some environmental problems. Tell your group what you think.

Some of the damage people are doing to the environment leads to climate change (e.g. global warming) and this leads to changes in the weather patterns in particular parts of the world. More or less rain falls. The winds are stronger or blow from a different direction. The dry season lasts for longer. The temperatures are higher. Such changes in the weather can have dramatic effects on the living things in a particular environment. This includes the people who live there and who are trying to produce food from the land. Rainfall, temperature and wind all affect the success of their crops. It may be that the changes happen because of the actions of people far away. This is one reason why the whole human race needs to act together to reduce and control environmental damage. No one is innocent and no one will be able to avoid the effects, one way or another.

Activity 10

You will need: the outcome of Activity 9, materials to produce a display or performance.

1. Discuss and plan with your group how you will present your ideas about your chosen environmental issue to the whole school, or to an audience from your community, e.g. parents.

2. Design the display or the props you need for the performance. If you choose to use music and drama, you will need to prepare them too. Make sure that your plans lead to a clear and interesting presentation.

3. Rehearse your presentation and make any necessary changes. Make sure that everyone knows what they have to do.

4. Present your topic to the audience and answer any questions that they may have about what you tell them.

The pictures show three things that children can do to take care of the local environment. Discuss them with your group and choose which one you would like to do for this term.

Activity 11

You will need: materials for the chosen task:
  (i)   compost heap – wood, plastic sheeting or wire netting, garden tools
  (ii) tree planting – young tree(s), spade, watering can or bucket, water
  (iii) litter collection – plastic bags, disposable gloves, rubbish bin or pit, spade
  (iv) paper and a pencil, books with information about compost heaps, tree planting or rubbish disposal.

1   Discuss with your group how you will carry out the activity. You should plan what you will need to do over the whole term, not just how you will start the process. Use the books to help you plan carefully.

2   Share out the tasks involved and begin the activity. Keep a record of the steps you follow and the results of what you do as you go along.

3   Continue with the process of caring for your local environment for the whole term. At the end of the time, write a report about the success or failure of what you have been doing.

4   Share your group's records and report with the rest of the class. Attend to what the other groups share with you about their projects.

There is no one answer to the problem of environmental damage, but there are four key ideas that can all help to improve the situation. They can be applied by all of us at the local level (at home and school and in the community), at the national level (by the government and businesses) and at the global level (by the United Nations and other groups of nations).

Reduction of the damage is certainly possible. We do not need to burn so much fuel in our vehicles – we can take the bus and not the car, we can use smaller cars, we can change to fuel without lead, etc. We can reduce the amount of pesticides we use on our crops. We can reduce the amount of paper and plastic we use in packaging. In many other ways, we can take better care of the resources of the planet.

One bus engine making smoke and 50 passengers on board.

Fifty car engines making smoke and 50 passengers on board.

Many of the things we throw away can be re-used. Clothes, vehicles, refrigerators, books, furniture – so many are just dumped because we are tired of them, even when there is nothing wrong with them. Such things can be re-used if we take the trouble to find new homes for them. This may be on a local level, where a neighbourhood scheme can collect and redistribute such things to those who need them. Nationally and globally there is a trade in valuable 'second-hand' machines and other things. Re-using is the second key idea to control the damage to the environment.

The compost heap is an example of the third key idea – recycling of materials. Organic matter, which was once part of plants or animals, can be turned into valuable fertiliser for plants. Making compost can be a small-scale thing we do at home, as well as a commercial activity by the local council. They can collect such organic waste and use it, often mixing the composted plant material with the end product from the sewage treatment works to make a commercial fertiliser that they can sell to farmers and gardeners. Around the world, councils and governments support the recycling of several different materials, especially those which are expensive to produce from raw materials, such as plastics, glass and metals. Paper is also recycled and this reduces the rate at which trees have to be cut down to make more paper.

Conservation is the fourth key idea – use everything with care and look after what we have. This is quite easy for even the youngest of us to do. We can turn off lights when they are not needed. This not only saves money on our electricity bills, it also reduces the amount of fuel that has to be burned at the power station. The same is true of taps. They are a wonderful tool, but they can waste a lot of water if we do not use them with care.

Conservation of the world's plants and animals is also our responsibility. This means controlling our disposal of waste and our use of the land and sea so that we do not destroy them or their habitats. Zoos used to be seen only as places of interest or entertainment. Now many of them are vital in preserving species that are endangered. Some have been able to breed enough of some species to set them free in their natural environment.

Hawaiian Goose

European Bison

These are just two of the species that have been saved from extinction by scientists in zoos. We can visit zoos and support them with our money, so that they can continue their conservation work. National parks and nature reserves are another form of conservation, set up to protect the habitats and all the living things found in them.

Glossary

Acid rain	rain polluted with gases from burning oil, gas and coal
Addictive	making the person dependent, unable to live without the substance
Anchorage	holding on to the soil that prevents the plant falling over
Anther	the part of the flower that makes pollen, the male sex cells
Auditory nerve	the nerve from the ear to the brain, which carries the impulses generated by sounds
Aurally challenged	unable to hear sounds across the normal range of pitch and volume
Bladder	the elastic bag in which urine is collected before being released from the body
Carbon cycle	the circulation of carbon from the atmosphere into living things and, after their death, back again
CFCs (chlorofluorocarbons)	chemicals used in refrigerators, aerosols, cleaning fluids and plastics. They damage the ozone layer and are banned in some countries
Communicable disease	also called infectious disease because the virus, bacteria or parasites causing the disease can be passed on (communicated) to another person in various ways
Cornea	the transparent 'window' that covers the front of the eye
Decibel (dB)	a measure of sound volume, the level of sound
Deforestation	cutting down all the trees in an area and leaving the land in danger of erosion
Drug	any substance, other than food, that causes changes in the body when it is swallowed, inhaled or applied to the body
Drug abuse	using drugs for the wrong purpose, or in doses that are excessive
Drug use	the use of drugs, either as medicines or for the effects they have on feelings
Ear drum	the thin membrane (skin) at the bottom of the ear canal, which vibrates when sound waves hit it
Egg	the female sex cell, made in the ovary
Endangered species	any plant or animal that is in danger of being made extinct
Environment	the surroundings in which organisms live, including the weather,

soil and competition with other organisms

Epidemic	an outbreak of a disease amongst people in a particular area, limited in area and time
Excretion	the process of removing waste products from the body. Includes urinating, sweating and breathing
Eyesight	the ability to see
Faeces	solid undigested food materials that are pushed out of the body through the anus
Fertilisation	the process of combining the male and female sex cells
Gaseous exchange	a process that occurs in the lungs, when oxygen is absorbed into the blood and carbon dioxide is removed from it
Greenhouse effect	the natural trapping of the sun's energy by the earth's atmosphere. Burning fossil fuels has increased this effect and global warming has resulted
Hearing aid	a device that helps a person to hear
Illuminated	lit by light from a source such as a candle or torch
Inner ear	the innermost part of the ear, where the vibrations heard as sounds are converted into nerve impulses and sent to the brain
Insect pollination	the transfer of pollen onto the stigma, which is done by insects visiting the flower
Intestine	the part of the digestive system where most of the processes of digestion and absorption take place
Iris	the coloured part of the eye, which can expand and contract to control the amount of light entering the eye
Joint	a point in the skeleton where two bones meet and are joined in a way that allows movement, e.g. knee, hip, jaw
Kidney	the excretory organ that removes urea and other materials from the blood, producing urine
Lens	the part of the eye that focuses the light entering the eye to produce an image on the retina
Lenses	the plural of lens
Liver	a large organ with many functions connected with nutrition, including the excretion of waste products into the intestine or the blood. It is above the stomach and below the lungs
Locomotion	movement from one place to another, e.g. walking, hopping

Luminous producing light, a light source

Middle ear the part of the ear containing the ear bones, connecting the ear drum to the inner ear

Mirage an image of water or an object in the distance, which is an illusion caused by the refraction of light passing through hot air

Mirror a surface that reflects an image of an object in front of it

Misperception when the sensation detected by the sense organ is wrongly interpreted by the brain, e.g. a mirage, the direction of a sound

Muscle tissue that can contract and relax. Bundles of this tissue are attached to bones and produce movement. Muscle is also present in other organs where movement is necessary, e.g. the heart, eyeball, digestive system

Non-luminous not producing light

Opaque light does not pass through

Optical illusion image that tricks the eyes, appearing different to what it is

Optic nerve the nerve that connects the eye to the brain and carries the impulses generated by light in the eye

Organ a part of a plant or animal that has a particular function, e.g. flower, brain

Outer ear the flaps of cartilage and skin on the sides of the head that help to pick up sounds, but have no sense of hearing

Ovaries a pair of small organs inside female animals and humans that make the eggs (the female sex cells)

Ovules the female sex cells made in the ovary of a plant

Ozone layer a layer of ozone gas high up in the atmosphere that protects the earth from the sun's very dangerous ultra-violet radiation

Pandemic a worldwide outbreak of an infectious disease, which continues for a long period over a wide area

Penis the male reproductive organ, which delivers the sperm inside the female vagina. It is also used for urination

Perception an awareness; detecting and interpreting information about the surroundings

Pesticides chemicals used to kill insects, weeds, fungi or other pests

Petals the part of the flower that is often colourful, large and attractive to insects, surrounding the reproductive organs

Pollen the male sex cells of flowering plants, made in the anthers

Puberty the period in the human life cycle when the reproductive system begins to function, i.e. eggs or sperm are first made

Pupil the hole at the front of the eyeball through which light enters the eye

Range of hearing the sounds, from the highest to the lowest pitch, that any particular animal can hear

Range of vision the distance over which any particular animal can see

Rectum the last part of the digestive system, where faeces collect before being pushed out through the anus

Reflection bouncing back. Light hitting a surface is more or less returned (reflected), or absorbed

Refraction bending light as it passes from one material to another, e.g. air to water, air to glass. It may lead to the separation of the colours in light to produce the rainbow effect

Reproduction the process in living things that produces new individuals

Reproductive concerned with or relating to reproduction

Retina the inner lining of the eyeball, which is sensitive to light

Root the part of the plant that absorbs water and minerals from the soil and anchors it

Scrotum a bag of skin between the legs of boys and men, which contains the testes

Sensory aid any device that supports the function of a sense organ to improve its performance, e.g. spectacles, hearing aid

Sepals the outer part of a flower. They are often green and cover the petals when the flower is closed in the bud

Sexual intercourse the process by which the penis delivers sperm inside the vagina

Shoot the part of the plant that connects the roots to the leaves, through which water, minerals and food travel

Skeleton the framework of bones on which the body is built. It allows movement and provides support and protection

Smog a mixture of fog, smoke and chemical fumes

Sound transmitters any object or material through which sound can travel

Sperm	the male sex cells, which are made in the testes
Stigma	the female sex organ in a flower, which has a sticky surface to trap the pollen grains
Sustainable development	using the natural environment in ways that do not destroy its resources, e.g. soil, water, air, living organisms
System	parts which are related to one another by their function, e.g. the nervous system consists of the brain, spinal cord, nerves and sense organs
Testes	a pair of small organs in male animals and humans which make the sperm. In humans they are housed in the scrotum
Toxic	poisonous
Translucent	some light passes through but we cannot see a clear image through such material
Transparent	light passes straight through and we can see through such material
Urea	a waste product that is disposed of in the urine that is produced by the kidneys
Urine	a pale yellow liquid excreted by the kidneys to remove waste products from the body
Uterus	the womb, the organ in female humans and other mammals where the baby grows and develops before birth. Part of the reproductive system
Vagina	the tube that connects the uterus to the vulva in female humans and other mammals. Part of the reproductive system. The penis delivers sperm into the vagina
Vision	the ability to see, sight
Visually challenged	unable to see normally over the range of distances and colours
Vulva	the external sexual organ in female humans and other mammals, the opening to the vagina
Water cycle	the continuous movement of water from the earth's surface into the atmosphere as water vapour and back again as rain, snow, hail and dew
Weather patterns	repeated features of elements of the weather in a particular area, e.g. rainfall, temperature
Wind pollination	the transfer of pollen from anthers to stigmas by the wind